Leaping Learners
Education, LLC

For more information and resources visit us at:
www.leapinglearnersed.com

Published by Leaping Learners Education, LLC, Verona, NJ 07044.

Every attempt has been made to credit each photo. Please contact us if there has been an error and we will resolve the issue.

Photo Credits

Monkey graphics © dreamblack46/stock.adobe.com

Cover, © Dr Morley Read/shutterstock.com; Title Pg. © Dr Morley Read/shutterstock.com; Pg. 1-2 © Dr Morley Read/shutterstock.com; Pg. 3-4 © max_776/stock.adobe.com; Pg. 7-8 © skynetphoto/shutterstock.com; Pg. 9-10 © skynetphoto/shutterstock.com; Pg. 11-12 © Patrick K. Campbell/shutterstock.com; Pg. 13-14 © Pavel Krasensky/shutterstock.com; Pg. 15-16 © KK Tan/shutterstock.com; Pg. 17-18 © By Sam DCruz/shutterstock.com; Pg. 19-20 © thatreec/shutterstock.com; Pg. 21-22 © SIN1980/shutterstock.com; 23-24 © Vladislav T. Jirousek/shutterstock.com, © Marek Velechovsky/shutterstock.com; Pg. 25-26 © *9999monkeys/reddit.com*; Pg. 28-29 © Dr Morley Read/shutterstock.com; Pg. 30 © aekikuis/shutterstock.com; Pg. 31 © Dirk Ercken/shutterstock.com, © Mary Long/shutterstock.com, © file404/shutterstock.com

All design by Sean Bulger
All other pictures by Sean Bulger

ISBN
978-1-948569-21-7

Dear Parents and Guardians,

Thank you for purchasing a *Matt Learns About* series book! After teaching students from kindergarten to second grade for more than seven years, I became frustrated by the lack of engaging books my students could read independently. To help my students engage with nonfiction topics, my wife and I decided to write nonfiction books for children. We hope to inspire young children to learn about the natural world.

Here at Leaping Learners Education, LLC, we have three main goals:

1. Spark young readers' curiosity about the natural world
2. Develop critical independent reading skills at an early age
3. Develop reading comprehension skills before and after reading

We hope your child enjoys learning with Matt. If you or your children are interested in a topic we have not written about yet, send us an email with your topic, and maybe your book will be next!

Thank you,

Sean Bulger, Ed.M

www.leapinglearnersed.com

Reading Suggestions:

Before reading this book, encourage your children to do a "picture walk," where they skim through the book and look at the pictures to help them think about what they already know about the topic. Thinking about what they already know helps children get excited about learning more facts and begin reading with some confidence.

Preview any new vocabulary words with your child. Key vocabulary words are found on the last few pages of the book. Have your children use each new phrase in their own words to see if they understand the definition.

After previewing the book, encourage your child to read the book independently more than once. After they have read it, ask them specific questions related to the information in the book. Encourage them to go back and reread the relevant section in the book to retrieve the answer in case they forgot the facts.

Finally, see if your child can complete the reading comprehension exercises at the end of the book to strengthen their understanding of the topic!

This book is best for ages 6-7
but. . .
Please be mindful that reading levels are a guide and vary depending on a child's skills and needs.

Matt Learns About . . . Army Ants

Written by Sean and Anicia Bulger

Table of Contents

Hi! My name is Matt. I love to discover and learn new things. In this book, we will learn about army ants. Let's go!

Introduction

Watch out!

You don’t want to get in the way of army ants!

Habitat

Where do army ants live?

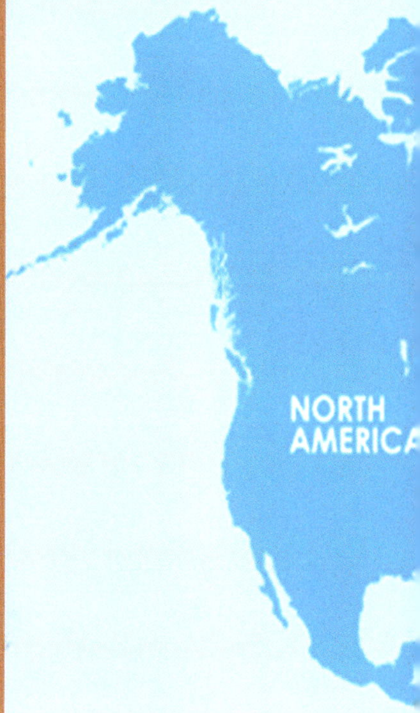
NORTH AMERICA

Army ants live in rainforests all over the world.

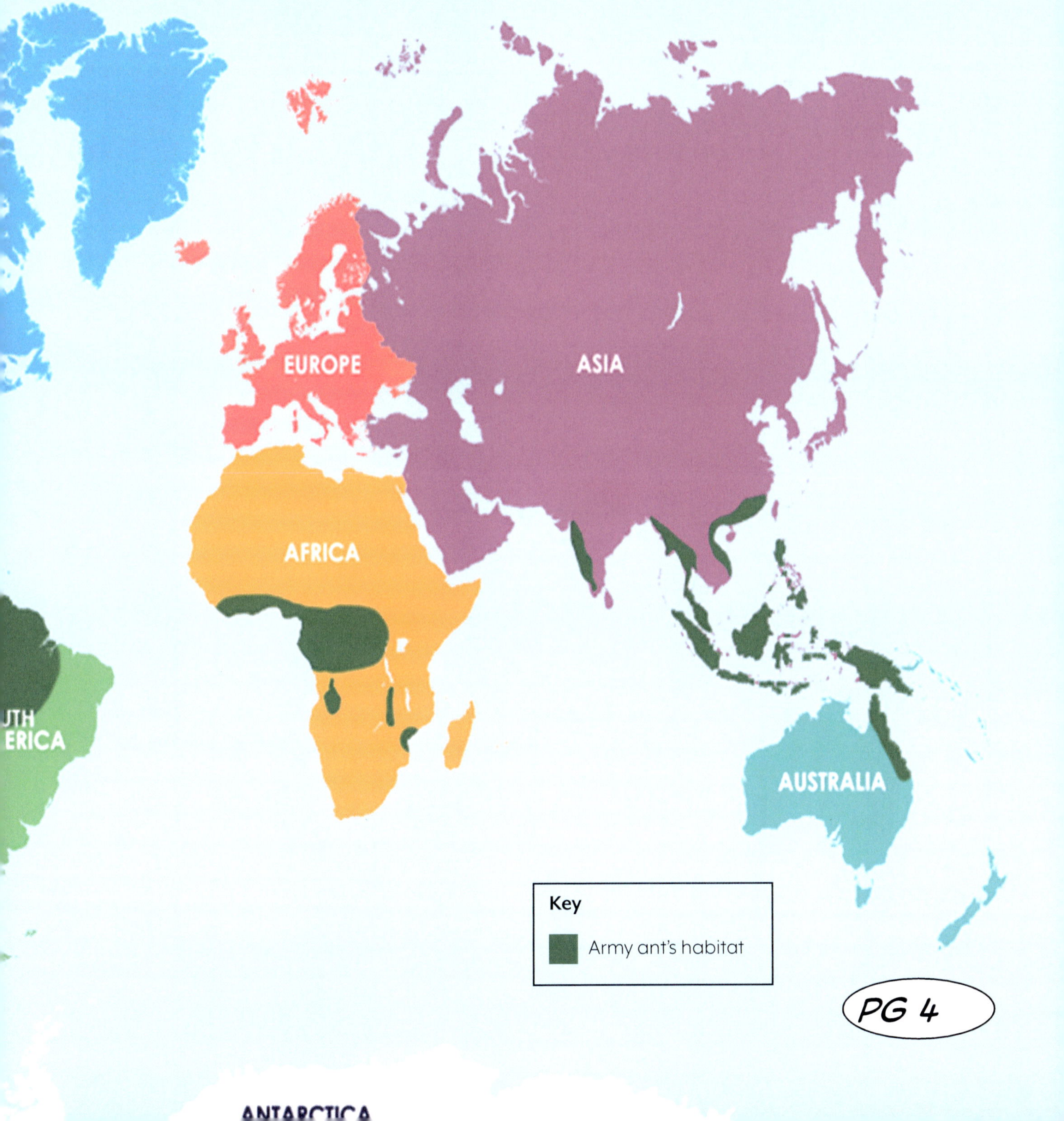

EUROPE
ASIA
AFRICA
UTH
ERICA
AUSTRALIA
Key
Army ant's habitat
ANTARCTICA

Army ants live in a **colony**. There are millions of ants in a colony.

The army ants join together to make their **nest**. The queen ant is safely in the middle of the nest.

Tricky word:
say. . .
col-o-ny

Body

What do army ants look like?

Army ants bodies have a head, thorax, and abdomen. They have six legs for walking and climbing.

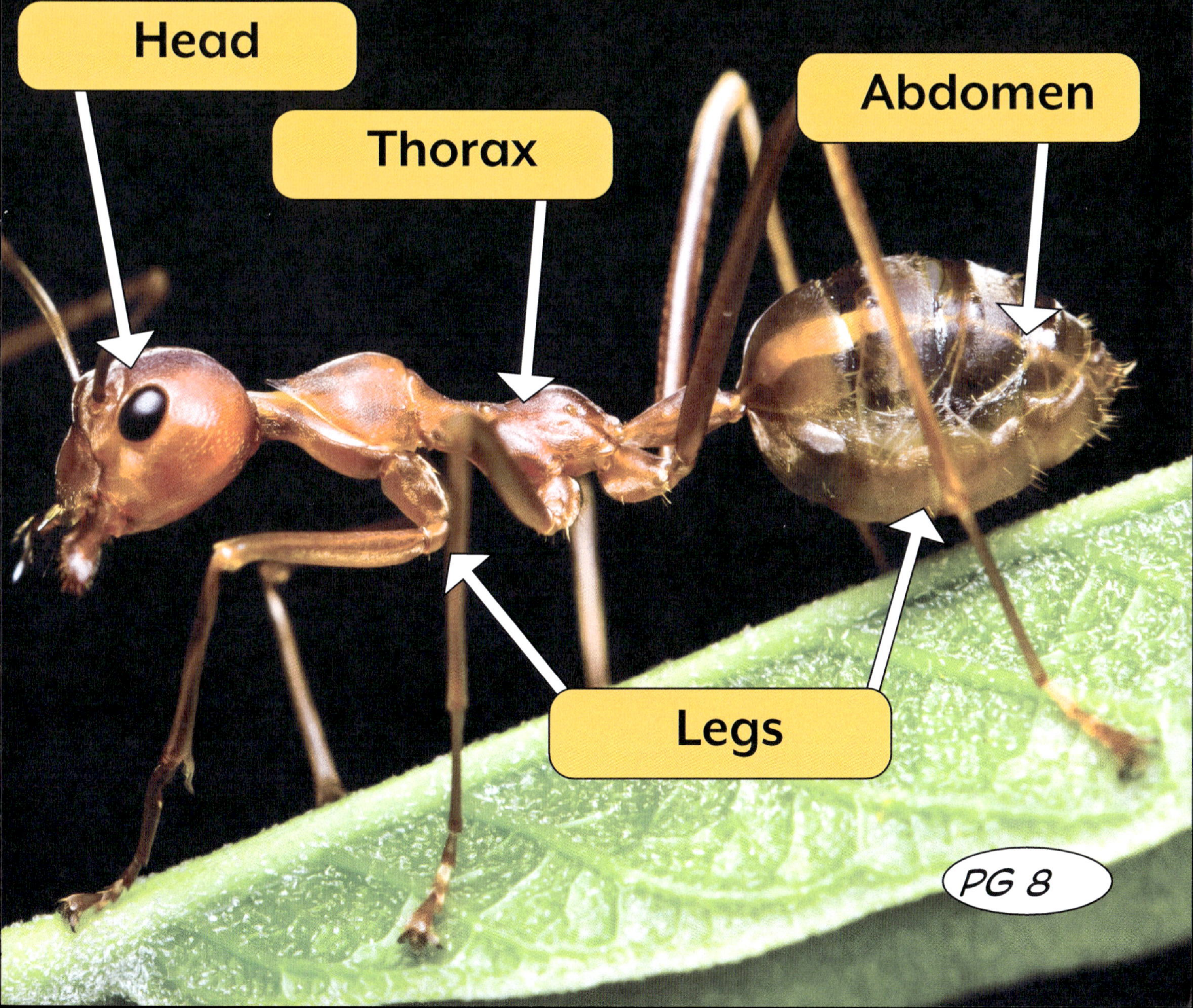
Head
Thorax
Abdomen
Legs

Worker Ants

Female ants are the workers and soldiers of the colonies. Female ants with small **mandibles** are worker ants. They carry food back to the nest for the queen.

Tricky word:
say...
man-di-bles
Mandibles

Soldier Ants

The largest female ants are soldier ants. They have extra-large mandibles. The mandibles are used to kill animals and protect the colony.

They cannot pick up anything with their large mandibles.

Male Ants

Male ants have wings. They use their wings to fly to other colonies in search of a new queen.

Food

What do army ants eat?

Army ants favorite foods are other insects, but they will eat anything they can kill.

Millions of army ants gather in a **swarm** to search for food.

Swarm

When the swarm sees a weak animal, it quickly attacks it.

When army ants attack, they sting the animal with **venom**. This makes the animal weak, and it dies.

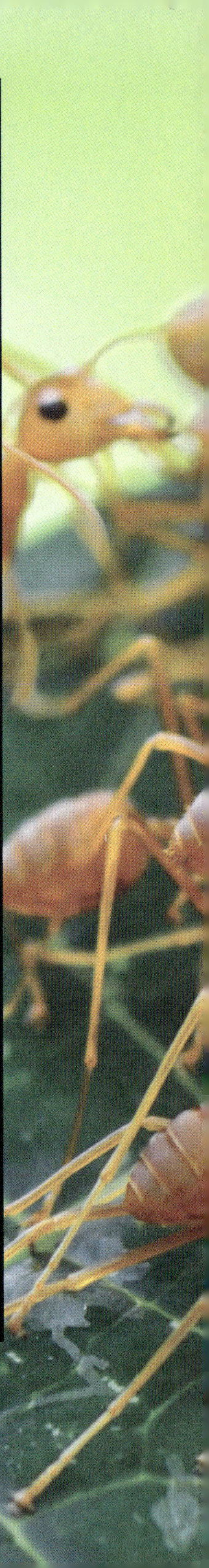

Team Work

Army ants work as a team. If the ants need to cross something while on the move, they will work together to make a bridge with their bodies!

Predators

Which animals eat army ants?

Even though army ants are dangerous, they still have **predators**.

Some of their predators are anteaters and spiders.

Laying Eggs

How do army ants have babies?

The queen ant can lay around 300,000 eggs in a few days.

The smallest female ants in the colony take care of the eggs in the nest.

Egg
PG 28

The life cycle of ants starts off as eggs. Second, they turn to **larvae**. Third, the larvae become **pupae**. Finally, they grow into their final form. Their final form is either a female worker, a male, or a queen.

Ant life cycle

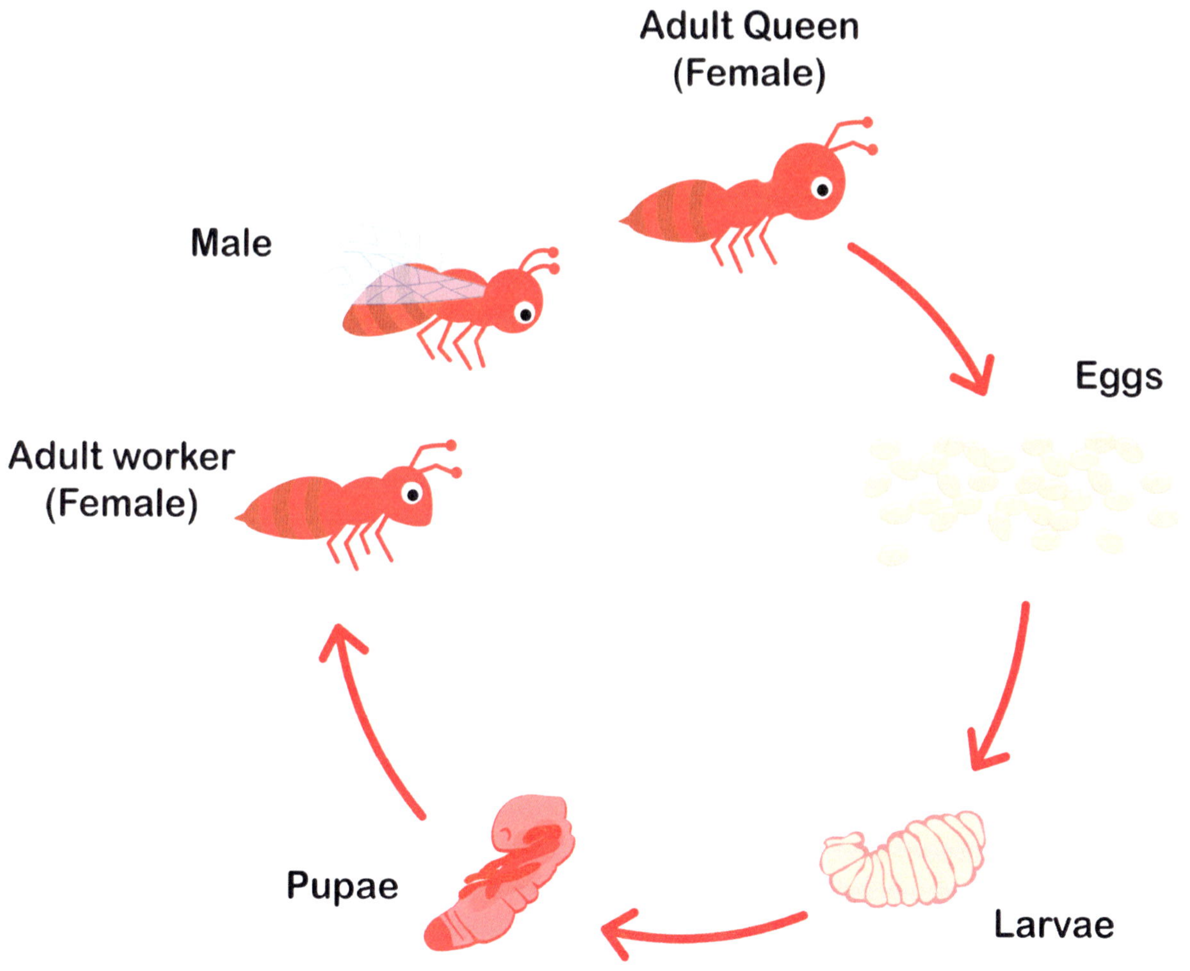

What are some interesting facts about army ants?

Army ants can lift 20 times their weight.

Army ants keep their nest very clean.

Army ants only live for about one year.

Glossary

A glossary tells the reader the meaning of important words.

Colony – Large community of ants

Nest – Home were the ant colony lives

Mandibles – Strong jaws of ants

Swarm – Large groups of moving ants

Venom – Poisonous liquid that ants shoot into other animals to kill or paralyze them

Predator – Animal that eats other animals

Larvae – Second stage of life cycle

Pupae – Third stage of life cycle

Draw a picture of an army ant.

Draw a picture of army ants eating.

The army ants are eating a

____________________.

Quiz

1. How do army ants make their colonies?

a. Using twigs

b. Using dirt

c. Using their bodies

2. What is the third stage of an army ant's life cycle?

a. Egg

b. Larvae

c. Pupae

3. How many eggs does a queen ant usually lay?

a. One

b. 300,000

c. 300

Common core standards:
RI. 1. 1 - Questions 1, 2
RI. 1. 2 - Question 3

4. What is one thing army ants eat?

a. Insects

b. Leaves

c. Bark

5. What is a swarm?

a. Many army ants together

b. Where army ants live

c. A liquid poison

6. What does the diagram on page 8 teach you?

a. What an army ant's body looks like

b. How army ants eat

c. Where army ants live

Common core standards:
RI. 1. 1 - Questions 4, 5
RI. 1.8 - Question 6

Want to learn about ocean animals? Check out the "Fay Learns About..." series!

Fay learns about . . .
Hammerhead sharks!
Reading comprehension strategies and activities inside!

Fay learns about . . .
Sea stars!
Reading comprehension strategies and activities inside!

Fay learns about . . .
Octopuses!
Reading comprehension strategies and activities inside!

Fay learns about . . .
Whales!
Reading comprehension strategies and activities inside!

Fay learns about . . .
Sea turtles!
Reading comprehension strategies and activities inside!

Fay learns about . . .
Hermit crabs!
Reading comprehension strategies and activities inside!

Fay learns about . . .
Seahorses!
Reading comprehension strategies and activities inside!

Fay learns about . . .
Clownfish!
Reading comprehension strategies and activities inside!

Fay learns about . . .
Jellyfish!
Reading comprehension strategies and activities inside!

Great for emerging readers ages 6-8

Want to learn about Farm Animals? Check out the "Katie Teaches you About..." Series!

Katie Teaches you about . . .
Ducks!
Reading comprehension strategies and activities inside!

Katie Teaches you about . . .
Horses!
Reading comprehension strategies and activities inside!

Katie Teaches you about . . .
Chickens!
Reading comprehension strategies and activities inside!

Katie Teaches you about . . .
Pigs!
Reading comprehension strategies and activities inside!

Katie Teaches you about . . .
Goats!
Reading comprehension strategies and activities inside!

Katie Teaches you about . . .
Cows!
Reading comprehension strategies and activities inside!

Great for early readers ages 4-6

Want to learn about colors? Check out the "Clayton Teaches You About..." Series!

Reading Level D
Clayton teaches you about.....
brown!
Hi! I'm Clayton. Let's learn about the color brown!
Reading comprehension strategies and activities inside!

Reading Level A
Clayton teaches you about.....
the color blue!
Hi! I'm Clayton. Let's learn about the color blue!
Reading comprehension strategies and activities inside!

Reading Level D
Clayton teaches you about.....
green!
Hi! I'm Clayton. Let's learn about the color green!
Reading comprehension strategies and activities inside!

Reading Level B
Clayton teaches you about.....
red!
Hi! I'm Clayton. Let's learn about the color red!
Reading comprehension strategies and activities inside!

Reading Level C
Clayton teaches you about.....
orange!
Hi! I'm Clayton. Let's learn about the color orange!

Reading Level C
Clayton teaches you about.....
yellow!
Hi! I'm Clayton. Let's learn about the color yellow!
Reading comprehension strategies and activities inside!

Great for early readers ages 4-6

Made in United States
Orlando, FL
20 March 2026

79517915R00029